Lee C. Corbett

A Study of the Effect of Incandescent Gas-Light on Plant Growth

Lee C. Corbett

A Study of the Effect of Incandescent Gas-Light on Plant Growth

ISBN/EAN: 9783337161583

Printed in Europe, USA, Canada, Australia, Japan

Cover: Foto ©berggeist007 / pixelio.de

More available books at **www.hansebooks.com**

WEST VIRGINIA

AGRICULTURAL EXPERIMENT STATION,

MORGANTOWN, W. VA.

BULLETIN 62.   OCTOBER, 1899.

A Study of the Effect of Incandescent Gas=light on Plant Growth.

By L. C. CORBETT.

—

[The Bulletins and Reports of this Station will be mailed free to any citizen of West Virginia upon written application. Address Director of Agricultural Experiment Station, Morgantown, W. Va.]

PLATE I.

Lettuce.

A STUDY OF THE EFFECT OF INCANDESCENT GAS LIGHT ON PLANT GROWTH.

The work of modern experimenters, while it may not be expressed in words, still indicates that the *kind* of artificial light used in physiological experiments upon plant growth, predetermines the results. If electricity is the source of light, affirmative results may be expected, such as those recorded by C. W. Siemens, in England; those of Deherain, at the Exposition Electric in 1871 at Paris, or Bailey and Rane, at a later date in our own country.

After having followed closely the work of the persons above mentioned and that of others along the same line, the writer became strongly of the opinion that it was not so much a question of the manner of making the artificial light, as the quality and intensity of the light.

The work of Deherain, of Paris, clearly shows that the quality of the light had much to do in determining the results. The naked arc electric was frequently hurtful while the same luminary enveloped in an opalescent shade or jacketed by a solution which absorbed certain portions of the spectrum, instead of being harmful really became somewhat of an advantage.

Bailey's work is a record of like results. The naked arc light being harmful at close range, while plants under a jacketed lamp were less injured and at given distances really benefitted by the use of the light.

Rane in working with the incandescent electric light records no such injury but shows quite as marked advantages as those recorded from the use of the arc lamp.

These results from such different sources would seem to show that it was a question of light of a given quality and intensity rather than a question of the source or manner of producing the light. In order to test this question and if possible to determine the range of the influence of an artificial light, the writer in the fall of 1895 began a test of the influence of the Welsbach incandescent gas light upon various plants growing in the green houses. This work modified from time to time to meet the requirements of the plant in use, has been continued up to the present, Aug. 1899.

In no case was the artificial light found to be a satisfactory substitute for daylight. But it is believed that could the conditions of the plants in the dark chamber during the day be kept as nearly normal as are the conditions for plants exposed to the artificial light at night only, the results would be quite different. The usual method is to place the plants to be exposed to artificial light only in a dark close box during the day and place them upon the benches under the light at night. This scheme while accomplishing the ends desired so far as the light factor is concerned, nevertheless places the plant under abnormal conditions during the time it is kept in the dark. M. Deherain in working upon this phase of the problem with the electric light had a miniature green-house constructed with darkened interior. That was much better than is usually the case, but even with plenty of air and room as would be the case in such a dark chamber, the results were not encouraging, and few plants were able to make more than a feeble growth with the electric light only.

In our own work little attention has been given to this particular problem. One experiment with lettuce being tried to demonstrate the influence of the gas lamp. This test was not satisfactory because the dark chamber for storing the plants during the day was too small and the plants damped off.

Since the completion of the first seasons work, which proved the stimulating influence of the incandescent gas lamps, the tests which have been carried on have been of an economic rather than a purely technical nature. True a number of scientific questions have been considered and answered so far as the conditions of these tests would admit of answering them, but in all cases the economic value of the work has been a consideration.

LETTUCE.

This crop has been used as the basis of all the light tests during the four seasons over which the work has extended. During that time nearly 10,000 plants have been used in the various tests. Varieties have played little part in this work.

The position of the lights have been altered so as to have first the north and then the south section of the house used as the light portion. In this way any conditions in one portion which tended to augment or detract from the true stimulus resulting from the light was counter-balanced when the reverse end of the house was used. The diagram of the ground plan of the house (Fig. 1) and the end view of the house (Fig. 2) will explain more fully what is meant. The position of curtain was constant, indicated in (Fig. 1,) but the lights, eight in number, were first arranged along the south end of house D, in the positions indicated by 1, 2, 3, 4, 5, 6, 7, 8, shown in (Fig. 1,) after having been used upon a crop of lettuce with section D' for the light house the lamps were then changed

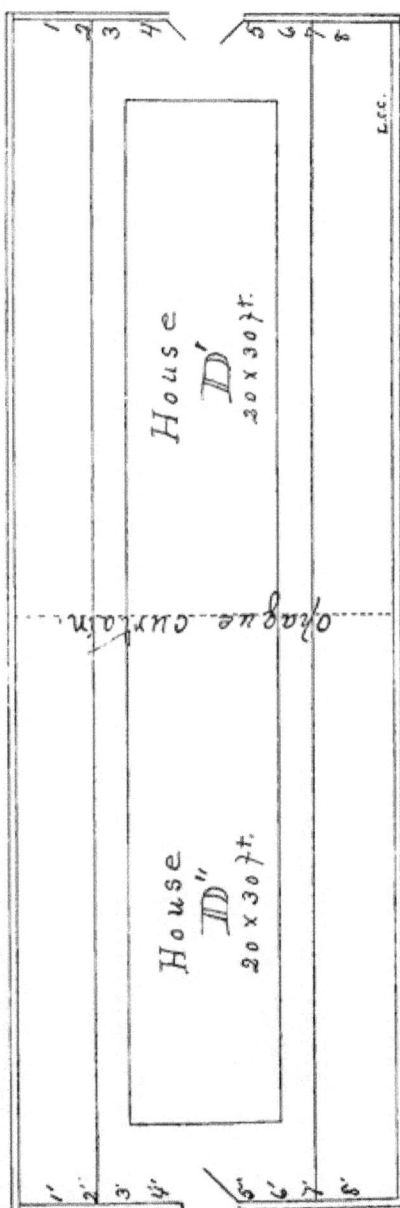

FIG. 1.

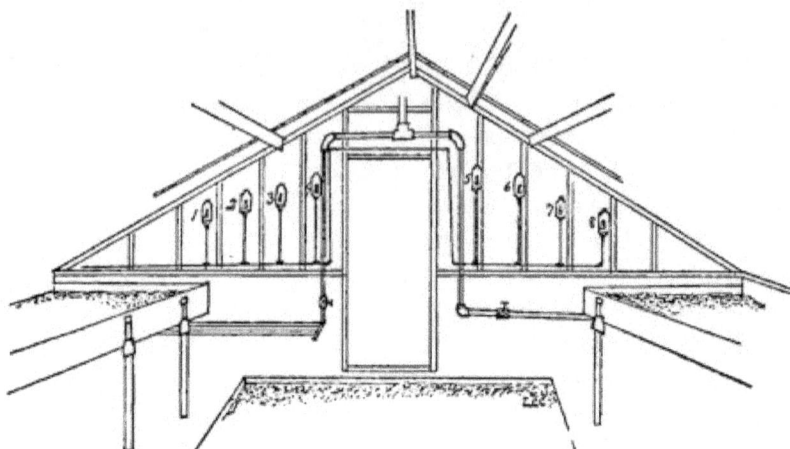

Fig. 2.

to the north end of the house where they occupied the position
1′, 2′, 3′, 4′, 5′. 6′. 7′. S , and that section of the house indicated as
D″, ῾.eu became the light section, section D′ then being the
normal. This alteration has been so made that D′ has been
the light section both during the fall and early winter and
spring portion of the year. The same is also true of section D″
when used as the light portion. In this way it is believed that
errors which are liable to result from differences in temperature
and light conditions, which are apt to occur in the most care-
fully constructed greenhouses, have been overcome. Then, too,

NORMAL.   Fig. 3.   LIGHT.

PLATE II.

FIG. 1.—Lettuce, Light House.

FIG. 2.—Lettuce, Normal House

the results presented are not those for a single season merely, but are the conclusions verified over and over again, in fact no less than twelve distinct crops of lettuce have been brought under the light conditions. In some cases the plants have not been subjected to the light until they were taken from thumb-pots and set 8x8 in. apart upon the permanent bench of the green house ; in other tests the seeds have been sown upon the benches and the young plants subjected to the light conditions from the moment they appeared above ground until fit for use. After a careful comparison of the two methods of treating the young plants the system of transplanting from pots and using the light upon the plant only during the period it occupied the permanent green house bench, was adopted as the more satisfactory commercial practice. The plants grown by the pot system responded readily to the influence of the light and gave a stockier and more salable product than could be grown by the other system—that of sowing the seed where the plant is to grow, even under the best of treatment. Lettuce plants subjected to the light without transplanting grow too tall without making a sufficient spread of leaves to give the greatest weight of product for the area occupied. The stimulating influence of the light was even more marked than in the case where pot-grown plants were used, but the commercial bearing of the light was a chief consideration, therefore, the extreme height growth which results from the light stimulus was sacrificed in the case of seed sown where plants were to stand for the more stocky and more symetrical growth obtained from plants grown by the pot method. Fig. 3 and Plate I illustrate the characteristic form and difference shown by plants grown by the pot method, one transplanted to the light section and the other to the normal section of the house.

Fig. 4 gives a view in the light section while Fig. 5 is its complement, taken at the same time in the normal section.

The photographs illustrate the striking difference in growth obtained from the use of the light upon plants grown by the pot system and transplanted to the bench, the plants being sub-

ject to the influence of the light at night only during the time
they occupied the green house bench.

The actual measurements of plants as well as their comparative growth is represented by the line charts. Chart (A) made
up from measurements taken Nov. 28, 1896, represents the

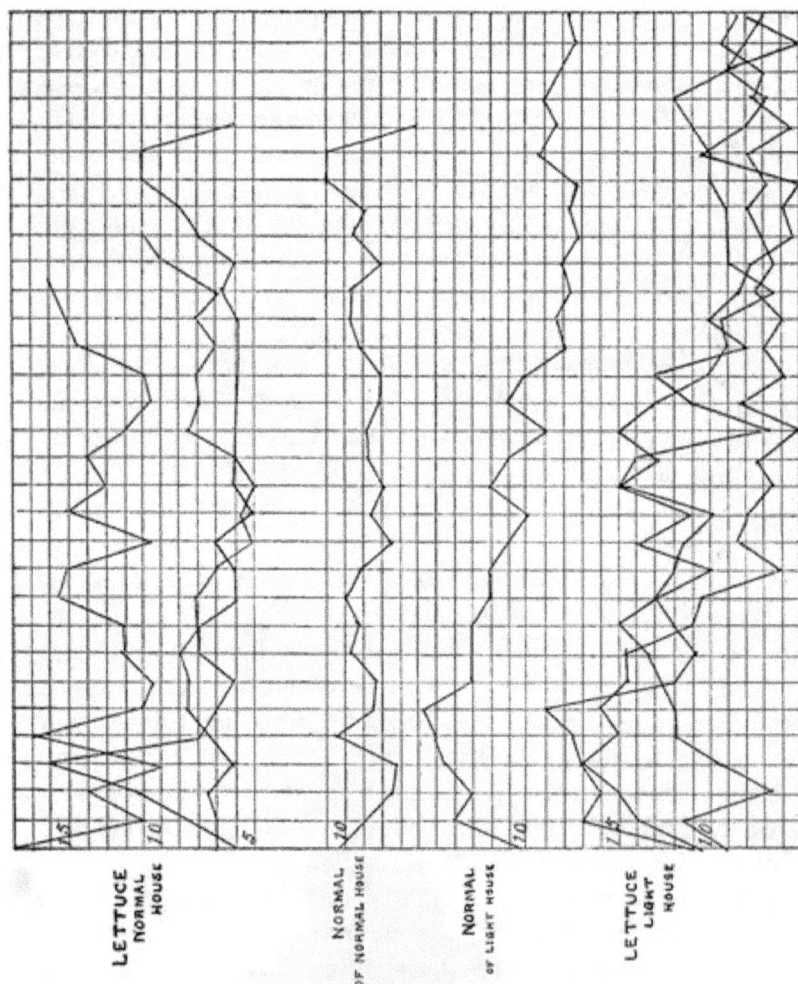

Chart A.

LETTUCE
NORMAL
HOUSE

NORMAL
OF NORMAL HOUSE

NORMAL
OF LIGHT HOUSE

LETTUCE
LIGHT
HOUSE

NOTE—The horizontal lines of the line chart are *one* inch apart, the numbers in the
spaces represent the number of inches above the base line, so that the height of any
plant in a row can be determined by locating the height of the first plant upon the
scale and counting above or below that point to get the others. The perpendicular
lines are eight inches apart.

growth of plants which had been retained upon the benches
much longer than is usual in commercial lettuce culture. In
fact it was the plan to show, if possible, which section of the
house would most quickly cause the plants to form seed stalks.
The curve of the light house is instructive in that it shows two
important factors in the behavior of the plants in the house at
this time It will be noted from a glance at the diagram that
the plants near the light, i. e. those represented at the left of
the curve, show a much greater growth than those near the
other end of the house. A less marked variation is shown in
the curve of growth of the plants in the normal section made
at the same time. In the case of the light house lettuce the
stimulus from the lamps is clearly shown at the left of the
curve. This house shows in a marked way the influence of the
lamp upon the plants near by as well as suggesting the range
of its influence.

The curve for the normal house made from measurements
taken Nov. 28, 1896, shows greater uniformity in the growth of
the plants from end to end. In order that these curves of actual measurements might be more easily compared and in order
that individual peculiarities of growth might be less apparent,
a line which is made up of the averages of the three lines of
measurements has been constructed and is marked *normal*.
This more clearly shows the general tendencies of the plants
under the two conditions. The normal for the light house is
very emphatic in its verdict for the stimulating influence resulting from the use of the light.

The results of the growth of plants which were allowed to
come only to a profitable commercial development is shown in
the line chart B.

A glance at these measurements and their normal repeats
the result above recorded, but in a less emphatic manner. The
line is sufficiently broken to clearly show the influence of the
light, as compared with the more regular line for the plants in
the normal house.

While the lines and the photographs tell the same story we

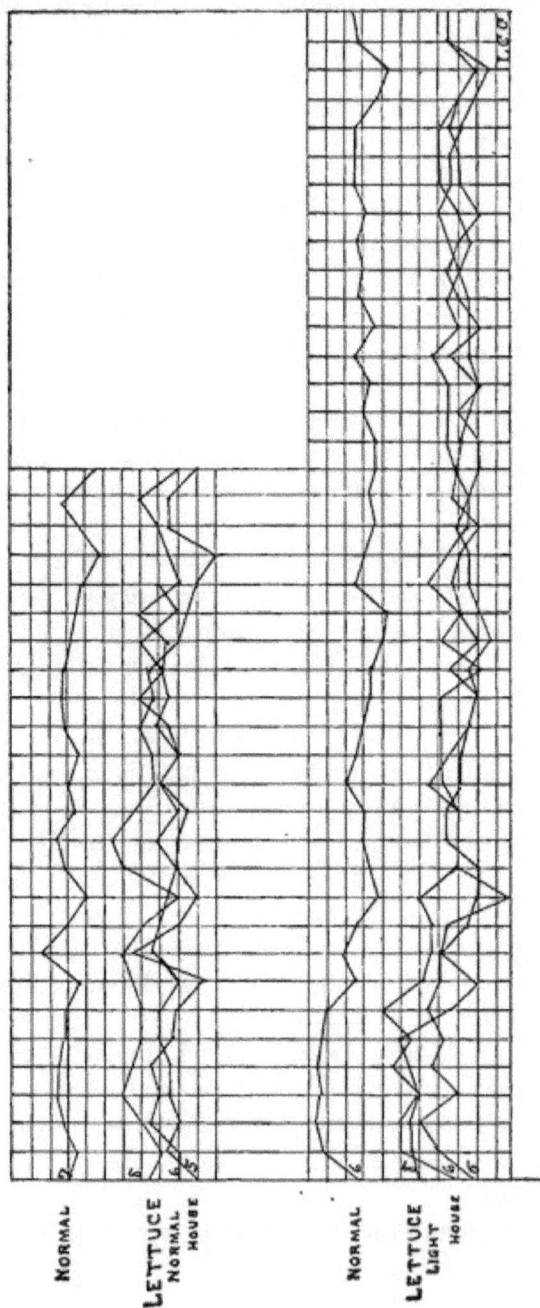

Chart B.

have it again set forth in the weights of the plants grown in the two divisions of the house.

The weight of 400 plants in the light area was 68.56 lbs. or .1714 lbs. per plant, in the normal 49.428 lbs. or .12357 lbs. per plant, a difference of 19.13 lbs. for the two sections and of .0478 lbs. per plant.

These plants were from seed sown Nov. 1, 1897, and set in the bed Jan. 6, 1898. The weights were made Feb. 21, 1898, The plants had, therefore, been under the influence of the light forty-six nights. During that period the lamps were turned on at five o'clock P. M., and extinguished at 7:15 next morning, thus making 655.5 hours that the plants were actually exposed to the stimulus of the light against 448.5 hours that they were in normal conditions. With the aid of the light the plants ex ceeded those in the normal section by 88.7 per cent. of the total weight of the plants in the normal house.

This clearly demonstrates that there was an actual gain in plant tissue in the lettuce under the influence of the light. Measurements and photographs show the difference in appear-ance, but weights alone are capable of proving that these seeming gains are real.

In fact it might be possible for a plant because of some exciting cause, to make a tall growth which photographs or measurements would show to be as striking as those above recorded ; yet the weight of the plant might be even less than that of one grown under normal conditions. Here, however, all three photographs, measurements and weights record the same results.

HOW PLANTS BEHAVE UNDER ARTIFICIAL LIGHT.

In connection with the foregoing studies auxanometer records were taken from a number of plants, both in the light and normal sections of the lettuce and spinach houses.

The greatly reduced scale of these records which was made necessary in order that they might be reproduced in the text has destroyed a great part of their usefulness. As originally made the record was twenty-four inches long. the growth of the plant magnified four times with the result that variations in

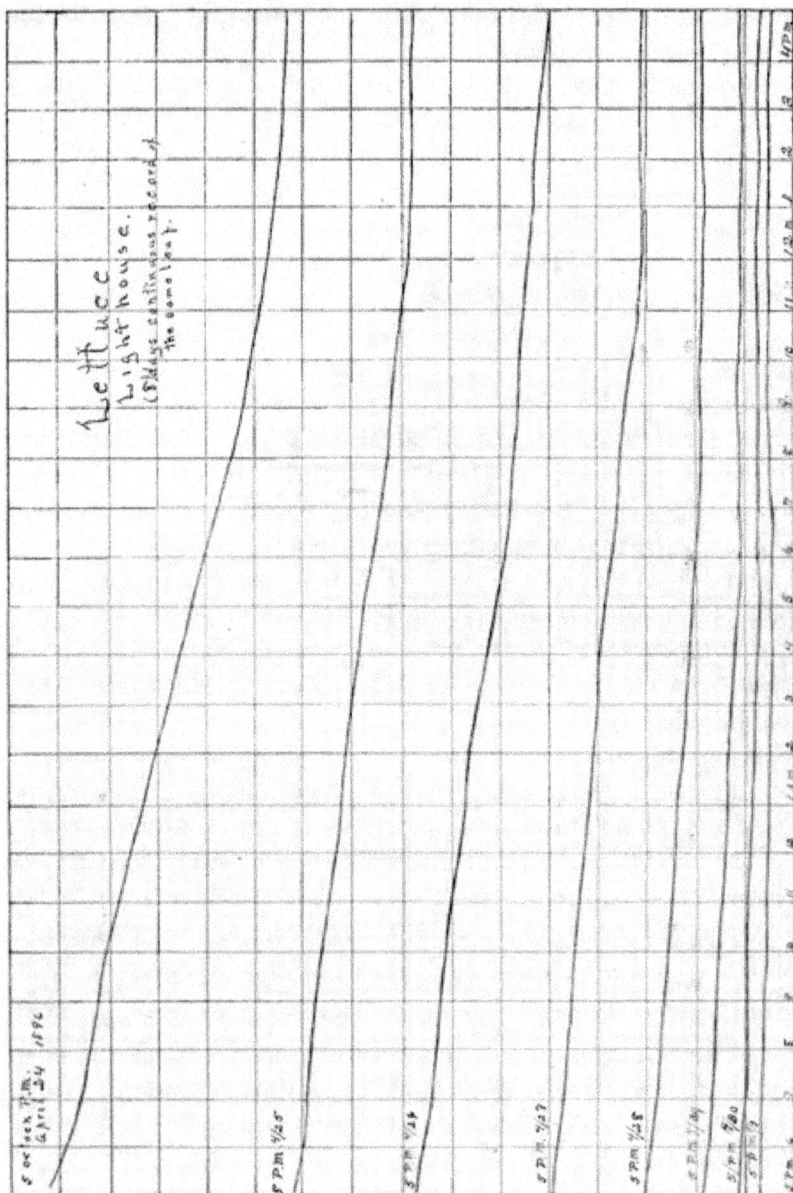

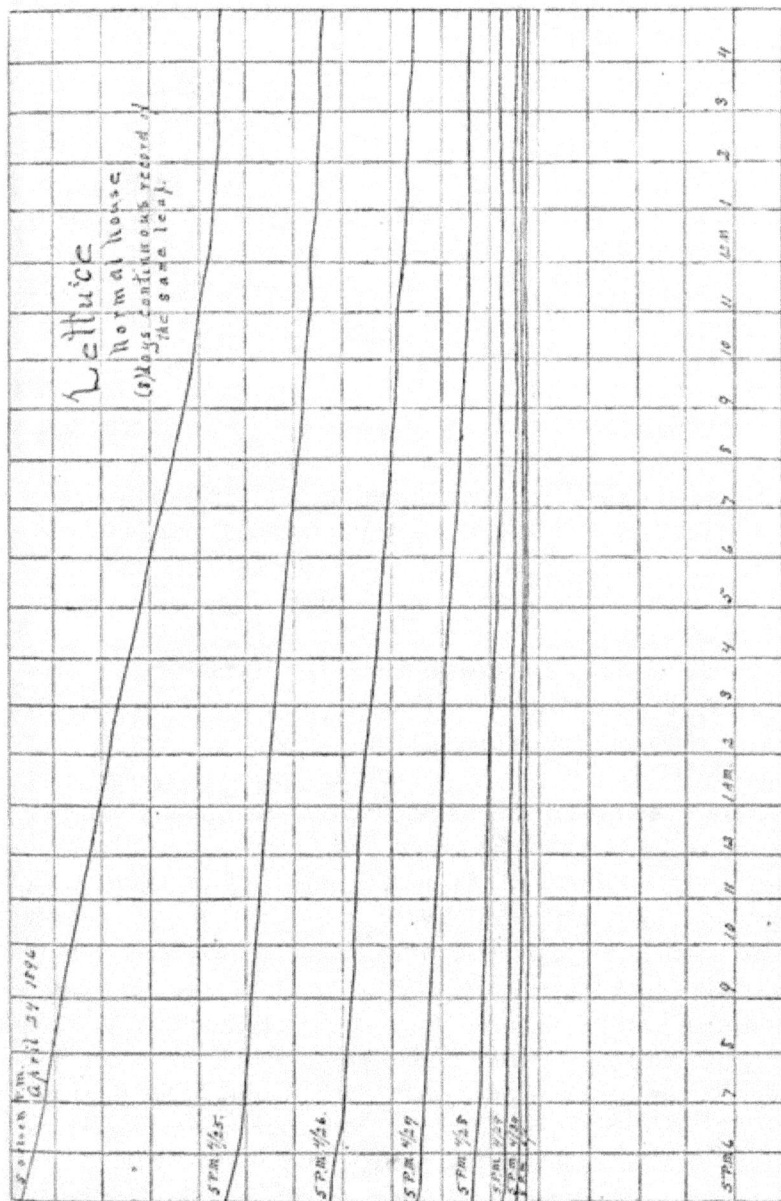

Lettuce
normal house
(8) days continuous record of
the same leaf.

Chart D.

the pitch of the curve were much more marked than appears
upon the reduced reproduction.

For that reason it has been thought wise to state the conclusions resulting from a study of the auxanometer records
without attempting to reproduce more than a sample sheet
from each the light and normal sections of the experiment.
The machine used for this work was one devised for experimental research by the writer and is shown in the accompanying illustration, (Plate III.) The simplicity and stability of the
device will at once suggest itself to the reader. A continuous
record from a single leaf of a plant in the light house is shown in
Chart C, while one of the same age and for the same period in
the normal house is shown in Chart D. These records read
from left to right, the fall of the line measures the upward
growth of the plant, i. e. the perpendicular distance from the
beginning of the first line to the beginning of the second, shows
the actual growth made by that leaf during 24 hours. or the
distance from left to right across the sheet represents 24 hours
of time.

A comparison of Chart C with Chart D will at once reveal the
more rapid rate of growth of the light house plants and will
also show the greater actual growth of the light plants during
the period of eight days over which this record extends. The
squares represent $\frac{1}{4}$ inch actual measurement; the growth of
the light plant it will be discovered covers some five spaces
more than the growth of the normal plant. The record then
shows that the light plant really grew one and one-fourth
inches more in eight days than the normal plant. The records
here reproduced represent the form in which the instrument
does its work and gives as well an idea of the character of the rate
and periodicity of growth of plants under the two conditions.
The records of plants grown under the stimulus of the light all
agree in showing that the fall of the recording needle was
greater than for plants in the normal house. Records from a
single leaf kept under continuous observation from the time it
was first unrolled until it had completed its growth show that
plants in the light house not only grow faster but that the

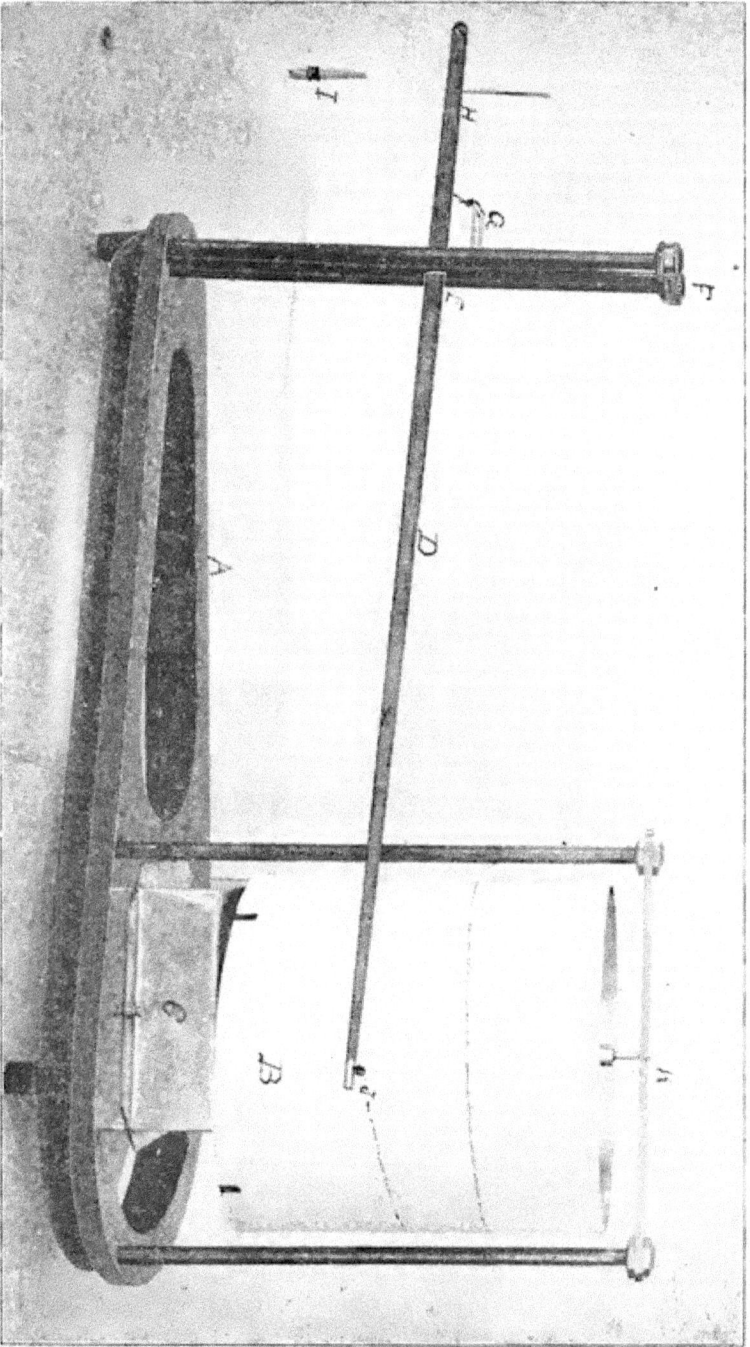

PLATE III.—Auxanometer.

PLATE IV.

FIG. 6.

FIG. 7.

period from appearance to full maturity is actually lessened for plants under the influence of the light.

PERIODICITY OF PLANT GROWTH MODIFIED BY THE INFLUENCE OF THE ARTIFICIAL LIGHT.

In this study no attempt is made to determine the periods of growth of various plants, it was taken up merely for the purpose of determining, if possible, whether or not the plants under the stimulus of the artificial light would show the same or a different period of growth from those in the normal house.

A comparison of the records from the two divisions of the house show that the period of most active growth is considerably longer in the case of plants under the influence of the artificial light than for those in the normal house. As a type illustration, the sheets here reproduced (Charts C and D) show that the most active period of growth for the plant in the light section began at *eleven* (11) P. M. and continued to *nine* (9) A. M., while during the same space of time the normal plant gave an ac'ive growth period beginning at four A. M. and continuing until (11) eleven A. M. In the case of the light plant the period was *ten* (10) hours while in the normal house the period was *seven* (7) hours. This then together with the greater rate of growth recorded in the light section easily explains the differences recorded in photographs, measurements and weights.

RADISHES.

In commercial lettuce culture it is a common practice to grow radishes between the rows, particularly when small lettuce plants are used for planting in beds of the greenhouse. In several instances during the course of the light tests radishes have been used as a "catch crop" with lettuce and in no case were any harmful effects noted from the combination. The plants here illustrated and upon which the conclusions here recorded are based were grown in the manner above described.

As is noted in the case of the sugar beet the radish shows a marked tendency to the formation of a large top when under the influence of the light; however, the increased top in the case of the radishes did not appear to be at the expense of the

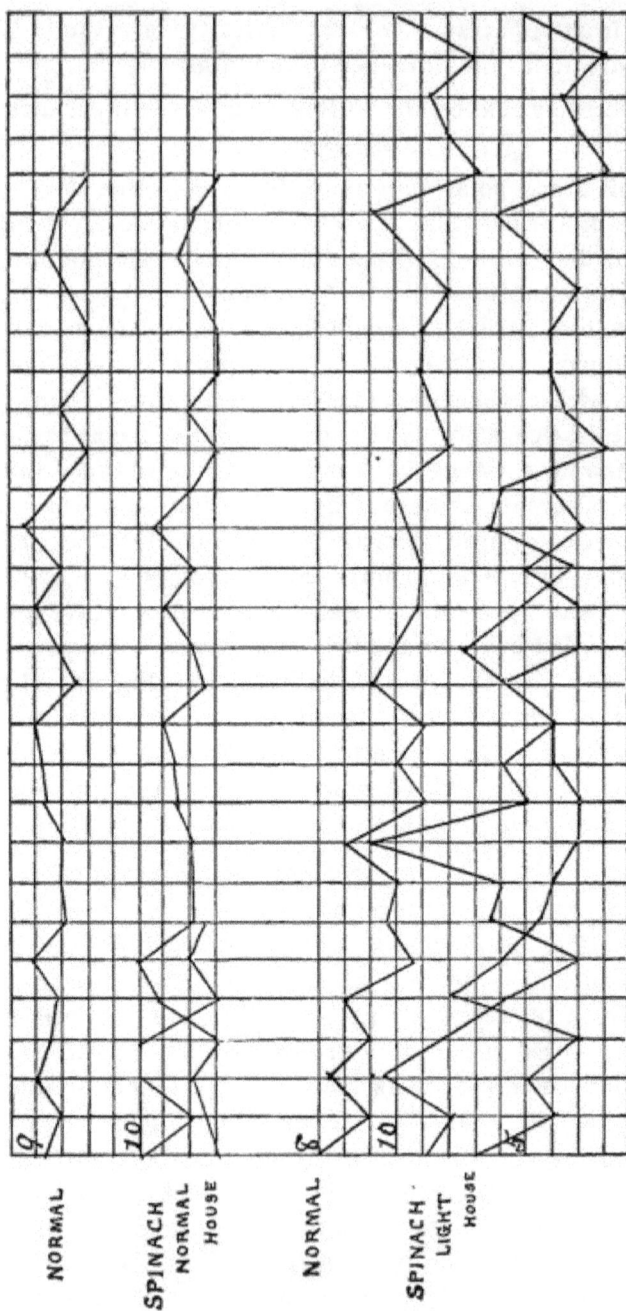

Chart E.

Fig. 8.

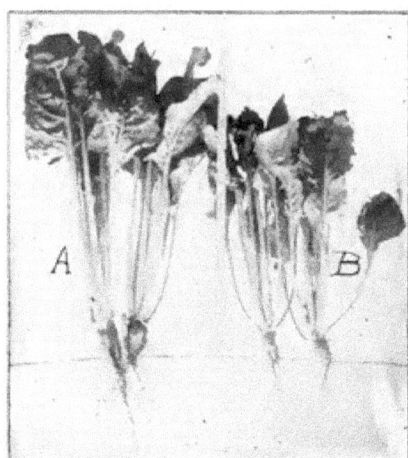

Fig. 11, p. 99.

root as is so apparent in the beet. In fact the roots of the light
house radishes were usually about the same size as those in the
normal house. The difference in the two being in favor of the
light house plants as is shown in the two turnip-shaped varieties illustrated by Fig. 6 and 7. The half-long variety shown
in Fig. 8, does not record a difference in favor of either the
light or normal section so far as root growth is concerned, while
the top growth of the light grown plant is considerably taller
than the other. These results, I am aware, are not in conformity with the results of Bailey with the use of the arc electric
light;[1] while the record of Rane's tests with the incandescent
electric light[2] during the season of '92-3 is indefinite.

Heliotropic effect of the incandescent gas light upon radishes
is noted, (Jan. 6, 1897), as being more marked than upon any
other class of plants in the test. At that time lettuce, spinach
and radishes were under observation. Later a slight leaning to-
wards the light was also noted in the case of lettuce, spinach,
tomatoes (all young seedling plants,) and quite as marked an
effect in the case of cabbage as was shown in radishes. It would
be difficult to show these variations either by photograph or
diagram, therefore a brief mention of the facts is all that is at-
tempted in this place.

SPINACH.

No plant used in these tests responded more to the stimulating influence of the gas light than spinach, although it was not
made a basis of any extended study because even if showing a
high susceptibility to such a stimulus its low commercial value
as a forced crop would not warrant the use of the light in growing it. From a physiological standpoint, however, the results
given by this plant are of interest and importance when compared with other nearly related plants.

The peculiar action of spinach when placed under the influence of the incandescent gas light is well illustrated by
Chart E, which not only gives the behavior of each plant as

1 Cornell University Exp. Station Bulletin No. 30.
2 Bulletin 37, W. Va. Exp. Station.

compared with every other in the test but also gives the actual measurements of the plant on a scale of 1-50 their actual size.

The platted curve for the light house spinach shows how greatly the plants near the lamp was stimulated by it. The curves in all cases begin with the plants nearest the light and record the behavior of each plant as distance from the lamp increases. In this way not only do we get the comparative influence of the light upon the plants at any given distance, but we have it for all distances covered by the length of the row in question, the result is that we have the range over which the light shows its influence. In few instances is this better shown than in the line chart for light house spinach. The first part of the curve shows a marked tendency to upward growth while as the other end of the row is reached much less tendency to height growth is shown and the behavior of the plants becomes more uniform. A glance at the curve of growth in the normal portion of the house shows a remarkable uniformity in growth of plants from one end of the line to the other. When it is remembered that these plants were grown upon the same green house bench from the same seed sowing, the two ends being set off only by an opaque curtain and one given the stimulus of the incandescent gas light as its only advantage over the other the difference recorded becomes more significent.

Not only was height growth stimulated in the manner above shown but a marked tendency to run to seed became noticeable in all plants near the light. This difference became so pronounced that the following note was made upon the behavior of a number of p ants; notes made Dec. 11, 1896:

First plant thirty-two inches from light, not healthy, small and with curled leaves.

Second plant thirty-six inches from light, has thrown up a seed stalk *three* inches high, plant healthy.

Third plant thirty-eight inches from light, seed stalk *four* inches high.

Fourth plant forty-two inches from light, seed stalk nine inches high.

Fifth plant forty-six inches from light, no seed stalk, plant unhealthy.

Sixth plant forty-nine inches from light, no seed stalk, plant unhealthy.

Seventh plant fifty-seven inches from light, seed stalk twelve inches high.

Eighth plant seventy-one inches from light, seed stalk seven inches high.

Ninth plant seventy-seven inches from light, seed stalk three and one-half inches high.

Tenth plant eighty-three inches from light, seed stalk four inches high.

Eleventh plant ninety inches from light, seed stalk four inches high.

Twelfth plant ninety-six inches from light, no seed stalk.

No other plant in either the light or normal house gave indication of the formation of a seed stalk at this time, this then in a manner measures the range of the greatest stimulus of the light upon the formation of seed stalks.

Note made Dec. 23, 1896. Two spinach plants in *light* house in bloom, and two others with well formed blossom buds nearly ready to burst into bloom. The interval of twelve days which has intervened since the last note was made is due to the plant which had the tallest blossom stalk and most advanced blossom buds having "damped off" since date measurements were taken, a more tardy plant then makes the date of bloom instead of the most advanced one. Up to this time, Dec. 23, no seed or blossom stalks have appeared in the normal house.

TOMATOES.

The plants used in the test were of three lots. Those which may, for convenience, be called Lot I, were normal seedling plants from seed sown Nov. 14. 1895; the young plants were pricked out on the 27th, placed in three-inch pots on Dec. 20 ; four-inch pots on Jan. 10, '96, and set in the greenhouse bench Jan 31, 1896, and began producing ripe fruits April 14.

Plants of Lot II were grown from cuttings made from healthy

plants which had borne a crop of fruit in the green house and
which had been grown from seed sown July 27, 1895. These
cuttings were made Nov. 13, 1895, and were subsequently
handled as follows: Placed in two-inch pots Dec. 12; three-
inch pots Dec. 20; four-inch Jan. 10, '96. and set in bed Jan.
31, same date as Lot I.

Plants of Lot III were grown from cuttings same as those of
Lot II, but the cuttings were made Dec. 5, '95. from plants of
the same seed sowing as those used for stock of Lot II. The
only difference between Lots II and III, being in the age of
of the parent plant when the cuttings were taken and in the
time of taking the cuttings. Those of Lot II being practically
one month older than those of Lot III, when set in the green-
house bench Jan. 31, 1896.

The accompanying table clearly brings out the behavior of
of these plants under the two conditions of the green-house, i.
e. in the light and normal areas. It must be stated in this connection that the plants here recorded as being in the light house
were at a distance of 20 feet from the light, and were only sep-
arated from those in the dark house by an opaque curtain
which was drawn each evening between two contiguous rows
of plants. The effect of the light upon the growth and product
of these plants was, therefore, less marked than upon plants
closer to the light. The record of plants close to the light and
those in a corresponding position in the normal or (dark) house
are here given simply for the suggestion it may offer, but not
for a basis of conclusion for the varieties are different in the
two cases, and might be sufficient to account for a part of the
difference.

TABLE I. TOMATO RECORD. LIGHT AND NORMAL.

Lot.	Variety.	Avg. No. Fruits per Plant.		Avg. Wt. Fruits per Plant.		Avg. Wt. individual Fruits.		History.
		Light.	Normal.	Light.	Normal.	Light.	Normal.	
1	Acme.	14.77	18.57	2.564	3.22	.216	.147	Seedling.
2	"	18.00	11.6	2.825	1.92	.156	.165	Cutt'gs 1 made
3	"	6.66	11.5	1.03	1.55	.180	.134	Cutt'gs 2 made
		38.43	40.68	0.444	6.69	.552	.446	

TABLE II. TOMATO RECORD.

Variety.	Avg. No. Fruits per Plant.		Avg. Wt. Fruits per Plant.		Avg. Wt. individual Fruits.		History.
	Light.	Normal.	Light.	Normal.	Light.	Normal.	
Liberty Bell. Beauty.	17.17	18.03	2.11	3.47	.123	.192	Seedling.
Liberty Bell. Beauty.	8.44	12.7	1.488	2.50	.170	.196	Cutt'gs 1 made
Liberty Bell. Beauty.	9.55	7.90	.922	1.42	.096	.179	Cutt'gs 2 made
	35.16	38.63	4.520	7.39	.395	.567	

From Table I of the tomato record it will be noticed that the totals for the normal plants exceed those for the light except in the case of the size of individual fruits. The fruits in the light exceeding in size those in the normal division. For the purposes of this study it is undoubtedly more just to use the seedling plants as a basis of comparison than to take even the average of the total product of the three lots.

If Lot I is decided upon the results become markedly pronounced and in every column marked *normal* we find a larger number than in the adjacent column marked *light*. The record of seedling plants in Table II gives the same verdict, and it is again repeated in columns of totals of Table II.

TOMATO BLOOM RECORD.

Variety.	History.	Date of First Bloom.	
		Light.	Normal.
Acme.	Seedling.	Jan. 31.	Feb. 11.
"	1st Cuttings.	Feb. 15.	Mar. 3.
"	2d Cuttings.	Mar. 15.	Mar. 23.

The above record gives a marked difference in earliness of bloom in favor of plants in the light area. This is in conformity with all similar observations upon spinach, lettuce, radishes, beets and flowering plants.

TOMATO FRUIT RIPENING RECORD.

Variety.	History.	Date of First Ripe Fruit.	
		Light.	Normal.
Acme.	Seedling.	April 14.	April 21.
"	1st Cuttings.	April 28.	April 23.
"	2d Cuttings.	May 5.	May 12.

With a single exception in the first made cuttings this table shows earlier fruits from the light house than from the normal.

PLATE V.

FIG. 9.—Sugar Beets, Light House.

FIG. 10.—Sugar Beets, Normal House.

SUGAR BEETS.

The sugar beet studies in the U. S. have clearly shown that certain areas, either because of atmospheric or soil conditions, are much better suited to the development of sacharine matter in the beet than others. In most cases the localities producing beets with greatest sugar content have been in high latitudes where short growing seasons are the rule and where the sun shines with great intensity for days uninterrupted by cloud or fog. These facts lead the writer to subject the sugar beet to the influence of the artificial light to see if it, like sunlight, would influence the growth or sugar content of the beet. Two deep beds were constructed in the greenhouses, one in the light house and another in an adjoining house where the light was not used. Great care was exercised to have all conditions except that of *light* alike. The soil for both beds was thoroughly worked over upon a mixing floor and was then divided into two lots, one being placed on the bench in the light house while the other half was put in the normal house.

Seed from a common lot was sown Mar. 27, 1899, in each bed. From the time of appearance above ground one lot was exposed continuously to the artificial light at night while the other was kept under normal conditions. Growth was good and on June 12, the plants of the light house were as shown in Fig. 9, and the normal, Fig. 10. While average individual plants from each bed showed the difference recorded in Fig. 11, in which "a" is normal and "b" light house.

On Aug. 28 the leaves of the plants had become badly affected with the "leaf spot"[1] and the product of each plant was harvested. Eleven plants from each plat were selected as types of their respective plats and photographed with the difference shown in Figs. 12 (Light house) and 13 (Normal house.) From these the 10 largest were selected for weight and analyses. The weight is recorded in the accompanying table which sufficiently explains itself:

---

[1] *Cercospora beticola*, Sacc.

WEIGHTS OF SUGAR BEETS.

Normal House.			Light House.		
Gross Wt.	Wt of Roots	Wt. of Tops	Gross Wt	Wt of Roots	Wt. of Tops.
8.5 lbs.	6.15 lbs.	1.35	4.775	2.0	2.775
Avg. .85 lbs	.615	.0135	.4775	.20	.2775

The average difference in weight of gross beet .3725 lb. in favor of normal conditions; in weight of roots .415 lb. in favor of normal, and .0425 lb. difference in weight of tops in favor of *light* grown plants.

The records in this table are certainly striking in the difference in weight shown between the two plats, in fact the difference was much more than was anticipated from conditions so nearly identical.

CHEMICAL ANALYSIS OF SUGAR BEETS

Lot.	No. Beets.	Total Wt. Uncapped	Total Wt. Capped.	Per cent. Sugar.	Per cent. Purity.
Normal.	10	99.0 grms.	62.2	6 69	5.53
Light.	10	31 2	19.5	8.35	6.10

While there was a marked increase in size and weight of root as shown both by photograph and weights, in the normal over the light house beets, yet a chemical determination of the sugar contents of the two samples shows very strikingly that the light influenced the sacharine matter in the root to a marked degree. This is in conformation of the theory advanced in the introduction to this experiment.

While growers who have given close attention to the behavior of beets in the northwest as compared with those in more easternly and southernly localities are fully aware of the differences shown by chemical analysis. I know of no other tests

PLATE VI.

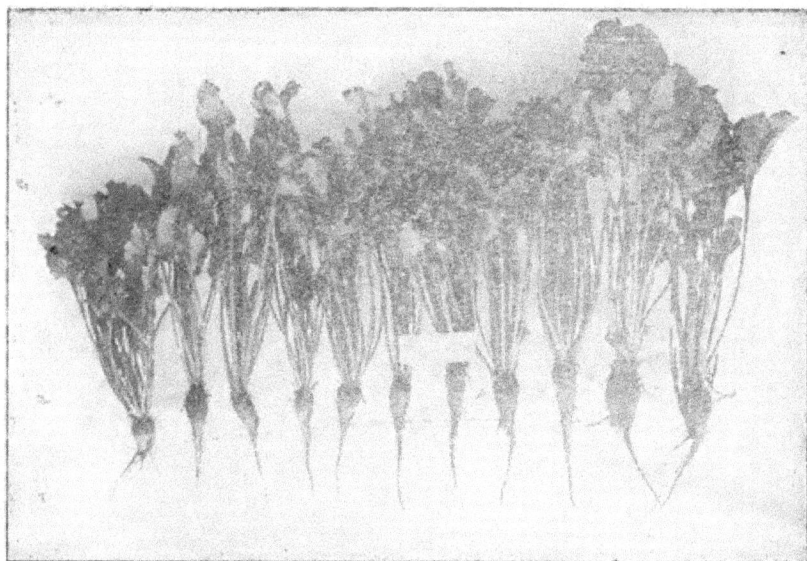

FIG. 12.—Sugar Beets, Light House.

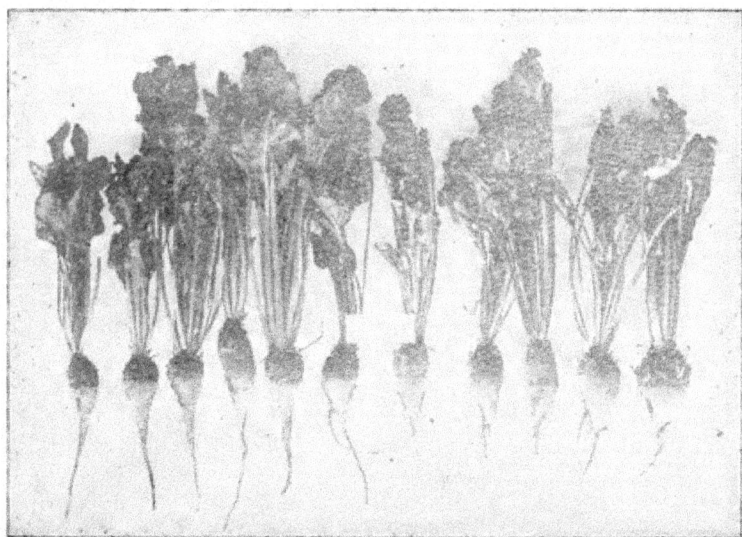

FIG. 13.—Sugar Beets, Normal House.

which have been carried out to show the influence of light upon sugar contents.

It is gratifying to be able to bring direct experimental proof in support of the light factor of the climatic conditions of the north-west in producing an increased sugar content of the beets grown in localities where intense sunlight prevails during a great part of the growing season. The results of this test emphasize the stimulating effect of the light on top-growth as compared with root-growth. Table of "weights," page (100), is emphatic on this point. In all respects save that of total weight of top, the normal house beets outweighed the light house plants, but in this respect the table is turned and the balance falls in favor of the light plants.

RANGE OF THE INFLUENCE OF THE INCANDESCENT GAS LIGHT AS SHOWN BY THE BEHAVIOR OF VARIOUS PLANTS.

Lettuce in this as in all other phases of the study takes the foremost places. In this test the conditions were modified by removing the opague curtain which had served to partition the house in all other tests. The whole length of the sixty-foot bench, by this change, came under the influence of the lamps. The record is that of plants grown from seed sown Nov. 17. 1898, potted Dec. 13 and set in greenhouse bench Dec. 28. 1898. Measurement being made Feb. 23. 1899. The actual growth of the plants of two rows, as well as their normal, is here recorded in line chart F. The influence of the light is less marked than in some other charts but is sufficiently pronounced, as shown in the *normal,*

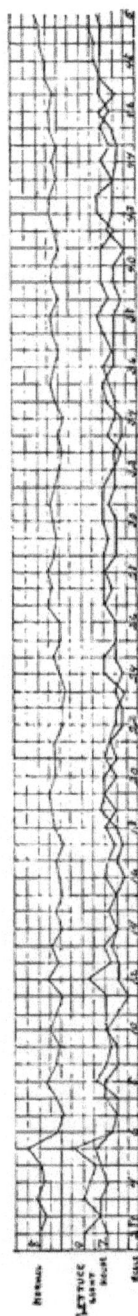

Chart F.

to give a basis for judging the range of the effect of the light.

The scale of feet recorded below the line of growth enables one to arrive at a just conclusion, regarding the influence of the light upon this crop of lettuce at a glance.

Beginning at the left hand and nearest the light, the plants standing 8 inches from one another and the first plants (18) eighteen inches from the lamp. As is shown in the *normal* it frequently, in fact almost uniformly, happens that the first few plants nearest the lamp are not as well developed as those farther away. A glance at the various line charts presented will show this clearly. Lettuce is, however, less markedly influenced in this manner than many other plants. This peculiar behavior is believed to be due to lack of light as illustrated by the diagram and not to injurious rays from it. The lamps were arranged as shown in Fig. 14. While plant No. 1 was 18 inches from the source of illumination the height of the lamp above the bed brought the plant really almost directly under the lamp. This is shown by the accompanying diagram.

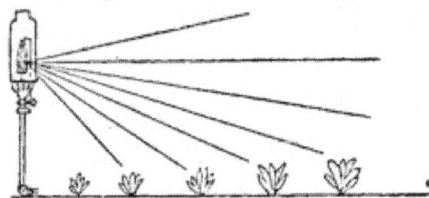

Fig. 14.

Here plants 1 and 2 get what light they receive from an unprotected mantle. The fact is, however, that plants 1 and 2 really receive very little direct light and are, therefore, quite like the plants referred to by Bailey as being in the shadow. Some light did really fall upon these plants and it was such rays as did not pass through the glass globe, but passed out between it and the frame of the lamp. The globes used being large, 4 inches in diameter, the space between the base of the mantle and the frame of the lamp was fully an inch which allowed considerable light to pass. Plants 1 and 2 which usually show a smaller growth than others in the same row may therefore be left out of consideration in measuring the range of the light. From the light there is a gradual gain in height up to the six foot mark ,then a gradual falling off takes place which reaches its lowest point at 24 feet, from that point on the line is

PLATE VII.

Fig. 16.—Cabbage.

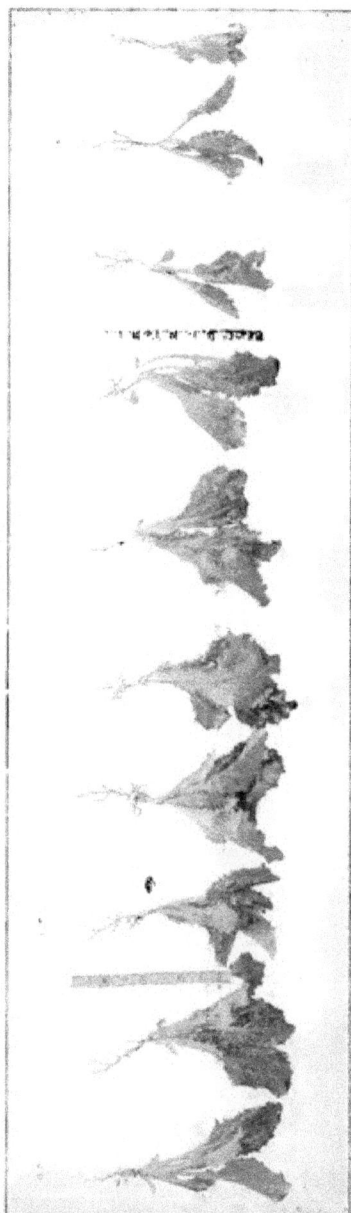

Fig. 15.—Lettuce.

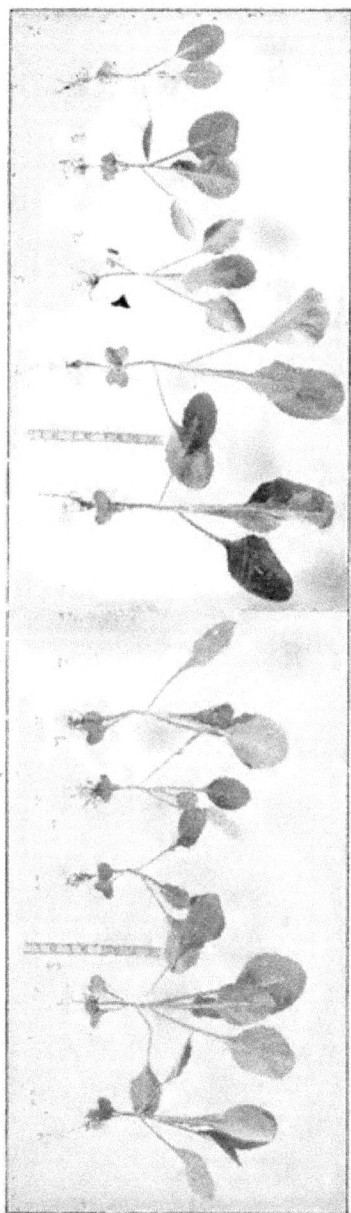

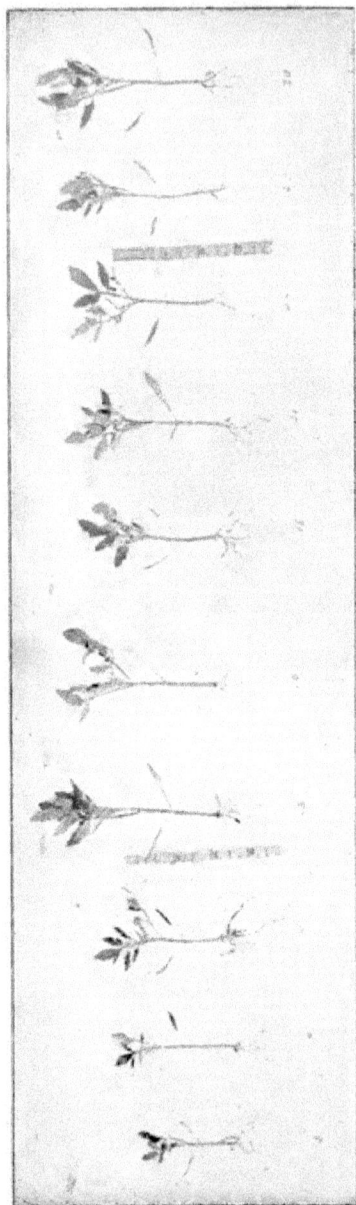

PLATE VIII.

Fig. 17.—Tomato.

Fig. 18.—Radish.

not unlike the variations of the lines shown by the normal house lettuce.

In the work which has been conducted, extending over four years, the range of the light seems to vary slightly for various plants, but the maximum is reached at from 12 to 16 feet from the lamp, while its influence is usually well marked at 24.[1] Beyond that limit little if any stimulating effect could be noted by any of the measurements used.

SEEDLING PLANTS.

In addition to the plants above described, seeds of lettuce, radishes, tomatoes and cabbage were sown in parallel rows in such a way that the range of the influence of the lamp might be studied. The results as shown by these plants are recorded in Figs. 15, 16, 17 and 18. The plants here shown were those which stood at 3, 6, 9, 12, 15, 18, etc., up to 30 feet from the light. The individuals chosen were such as happened to stand at that particular point in a row of seedlings : there is therefore a less uniform graduation from point to point than actually shown by the plants upon the bench. The photographs are all of one accord, however, in showing the marked effect upon the plants up to 12 or 15 feet, as the case may be, and then follows a more or less uniform series of plants, as is shown in the case of the tomato, Fig. (17); or a gradual diminution in size as is shown in the lettuce, Fig. (15); the radishes, Fig. (18); and in the cabbage, Fig. (16.)

The striking results of the stimulating influence of the light is shown by none of the plants used in the tests better than by the spinach and cabbage ; the radish is third in susceptibility, lettuce fourth and tomatoes fifth.

The results recorded by the behavior of the plants is interesting when considered in connection with the natural distribution of the luminary rays from the incandescent mantle of the Welsbach burner.

The way in which luminous rays are given off by the electric arc lamp and by the Welsbach incandescent gas lamp is at

once striking and interesting. The influence of the form of
the luminous centre is clearly shown in the accompanying
diagrams, (Figs. 19 and 20.) The diagrams show how the light
is thrown off by the bare lamps, Fig. 19 showing the light distribution for the arc electric, and Fig. 20 that of the incandescent gas lamp of the Welsbach Co, "By a bare Welsbach
lamp is meant the Welsbach lamp without globe or reflector.

TABLE III.

LIGHT DISTRIBUTION OF THE BARE WELSBACH GAS LAMP.

Angles measured from the horizontal.	Below Horizontal.	Above Horizontal.
Horizontal 0°	100	100
10°	96.2	99.7
20°	88.4	98.3
30°	77.0	90.5
40°	63 3	82.7
45°	58 6	75.4
50°	50.6	70.9
60°	32.4	58.8
70°	14.3	42.4
80°	3.5	30.0
9.°	0.4	

Table III shows that there is a continous decrease in the luminous intensity as we pass into the different directions lying below the horizontal plane of the lamp, and this decrease is more
rapid the more nearly we come vertically under the lamp.
Vertically under the lamp it is almost dark, for there the intensity falls to only 0.4 per cent. of what it is at the same distance
horizontally.   Above the horizontal plane the distribution of
light is very similar, but the decrease is far smaller.   The results show that the greater part of the light produced is sent
upwards to about 20° above the horizontal plane.

The rapid diminution of lighting power as we come underneath the We'sbach lamp, which is peculiar to this lamp, is not
astonishing when we consider that the mantles, after a very
short time, assume a form which is smaller above, and that any
light sent from the upper portions of the mantle is obstructed

PLATE IX

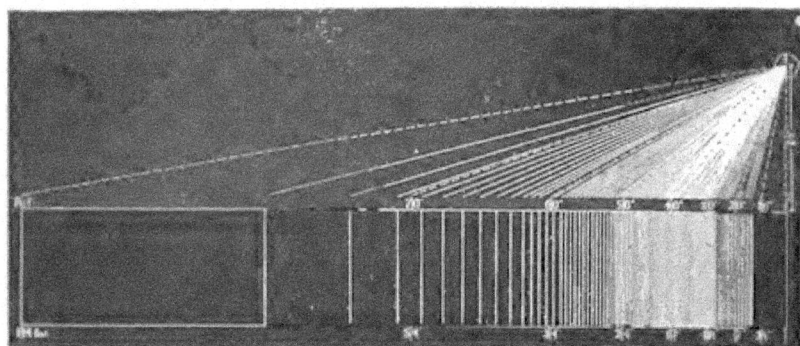

FIG. 19.—Bare arc lamp.

FIG. 20.—Bare Welsbach.

in its path downwards by the wider lower portions. There are similar phenomena, however, even in the Argand flame."

A comparison of the diagrams shown in Figs. 19 and 20 gives a clear idea of the distribution and range of the light from the arc electric and from the Welsbach gas lamps. The distance between lines gives an idea of the intensity of the lights at various angles and distances. Angles are noted at the top of the scale bar and distances underneath. The angle covered by the gas lamp is much greater than for the electric light ; the distribution of the light from the electric lamp is therefore more localized and is more intense than is the case with the Welsbach. The greatest quantity of light is given off from the electric light within the area marked 11½ to 24 feet, from the perpendicular ; while the light is much less intense from the Welsbach and is given off over a belt covered by 8½ to 17 feet, from the perpendicular, 12 feet being the place of maximum intensity of the light.

It is a strange but striking coincidence that the diagram here given for the Welsbach lamp should so closely correspond with the growth of the seedling plants reproduced in Figs. 15, 16, 17 and 18.

The light diagrams, Figs. 19 and 20, are from the Journal of the Franklin Institute, and are here used by courtesy of the publishers.

The range of the Welsbach lamp, as stated in the results of the test, is from 12-16 feet for the greatest stimulating influence with a marked influence up to 24 feet. The diagram makes this deduction even more emphatic and gives a reason for it, and what is still more gratifying, the conclusions and deductions recorded in the description of the experiments were all made before the writer had any knowledge of the light chart here reproduced. The peculiar behavior of the plants nearest the lamp now becomes easily accounted for. The fact is that plants in this position were in the dark and received no stimulating influences from the light, instead of being injured by it.

THE CHARACTER AND QUALITY OF THE LIGHTS USED.

Line chart G gives the plotted curve of the spectra for the arc electric light, for the Welsbach light and for the incandescent electric light in terms of sunlight, as given in Table IV.[1]

TABLE IV.

Wave Length. millionth of a millimeter.	Color of Spectrum.	When Sunlight is unity the lamps bear the following ratio to it.		
		Incan. elec.	Arc-electric	Welsbach.
800	Ultra-red	11.30	1.68	2.94
760	Red	4.90	1. 7	1.88
686	Red...	2.69	1.29	1 61
656	Orange	1.25	.98	1.09
589	Yellow	1.00	1 00	1.00
570	Yellow	.80	.82	.84
542	Yellowish Green	.48	.71	.58
527	Green	.38	.77	.47
517	Green	.30	.77	.39
500	Greenish Blue	.20	.57	.23
486	Brilliant Blue	.17	.56	.17
466	Brilliant Blue.............	.45	.58	.15
458	Brilliant Blue	.14	.61	.13
447	Dull Blue	.13	.76	.11
431	Dull Blue	.10	.83	.08
423	Violet	.08	.92	.06
416	Violet	.06	1.12	.04

The lettered line, A to G, represents the spectra of sun-light, while the other curves give the ratio which the various colors in the lamps bear to sun-light. The letters A to G represent the primary colors of the spectrum. A is red, B orange, C yellow, D green, E blue, F indigo, G violet.

Lines which extend above line A–G have more of the colors through whose territory they pass than does sun-light, i. e. all three lamps have more red, orange and yellow than sun-light when their sodium or D lines are made equal. The lines which fall below line A–G show the light to be poorer than sun-light in those colors through whose territory they pass, thus all three lights are poorer than sun-light in blue while while the arc-light alone contains as many violet rays as sun-light.

It will be noted that the lights stand in the following order

[1] See an article in "Progressive Age" of July 1. 1899, written by Truchot. The tests were made by Von Mutzel on a Glaw-Wogel photometer.

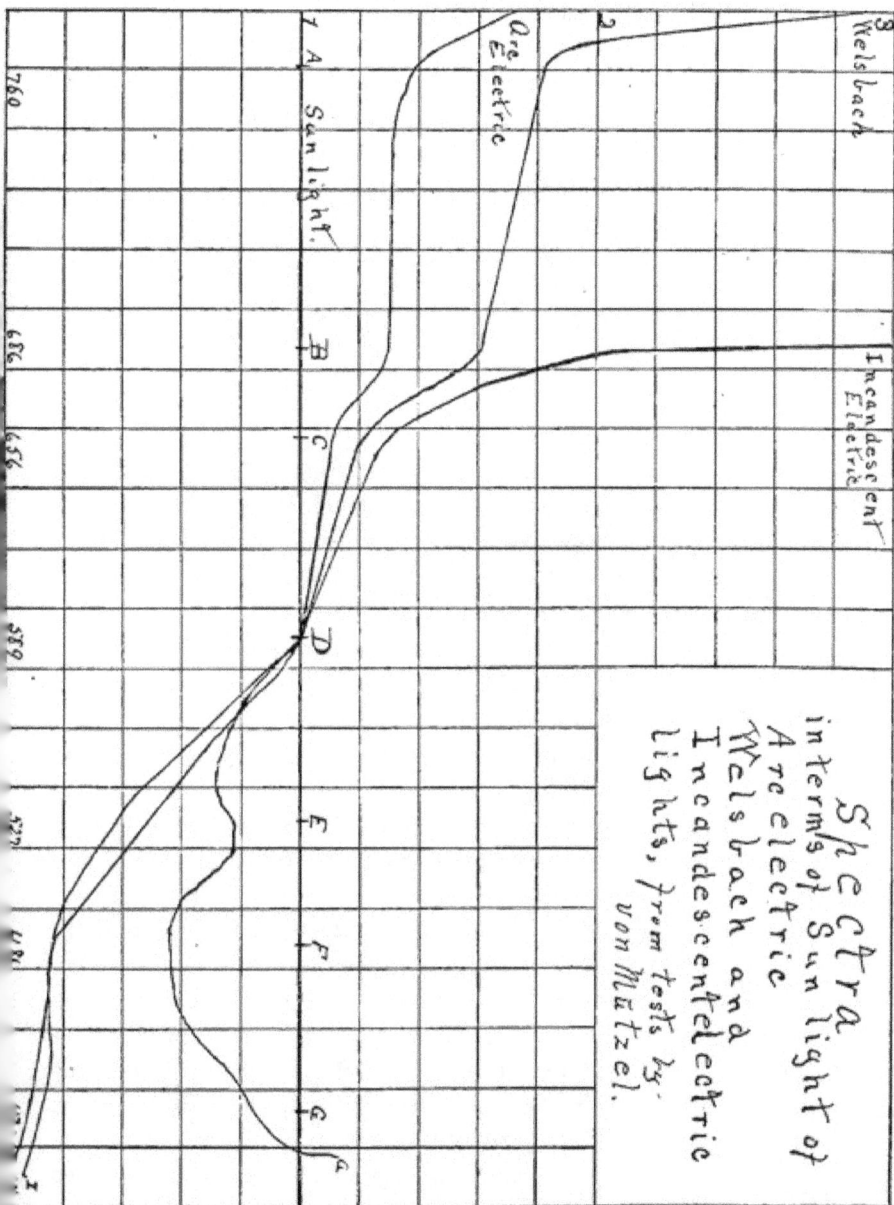

Chart G.

in regard to red, orange and yellow—incandescent electric most, Welsbach next, arc-electric least or nearest to sun-light. The arc-electric, while poorer in reds and yellows, is much richer than the Welsbach in blue, indigo and violet. The Welsbach has more blue than the incandescent electric, yet they are equal in the indigo with the Welsbach slightly weaken in violet and ultra-violet.

With this comparison of the three lights which have been used for the purpose of stimulating growth in plants some of the peculiar results obtained may possibly become less mysterious.

The work of Camille Flammarion[1] with plants in green houses constructed with glass of different colors, and the remarkable results which followed from the use of red, as compared with blue, green and white or normal light, which is in the main a coroboration of the work of Bailey and the writer published in Bulletin 55, of Cornell University Experiment Station, together with the spectra of the light used in such cases gives a possible basis of explaining certain results obtained by Bailey with the arc light, by Rane with the incandescent electric and by the writer with the Welsbach.

The three lights which have been successfully used for the purpose of stimulating plant growth are quite different from sunlight and from each other in quality, yet each investigator reports more or less marked results from the influence of the light.

Bailey used a light of high candle power, containing little more red than normal light, but also weaker in blue and again richer in ultra violet rays. Rane used a light of exceedingly low candle power (60 to 80) and very rich in red, orange and yellow, but poorer than the arc or Welsbach in blue, but equal to the latter in indigo or bright blue, and not different from it in violet and ultra-violet.

Bailey reports injury from the use of the arc light as does also Dherrain, while Rane records no such effect and with the

1 Translated for Experiment Station Record, Vol. X, No. 2.

Welsbach of 560 candle power in our own experiments no injury could be attributed to the light.

From all these results it would seem that the stimulus following from the use of the Incandescent Electric and the Wellsbach incandescent gas light might be explained, on the basis of the low candle power used, from their richness in red and orange rays, while the injury from the arc light may, as Prof. Baily remarks, be attributed to the large proportion of the active chemical rays—ultra-violet rays in the arc-electric.

TEACHINGS OF THE TESTS.

1. The incandescent gas light of the Welsbach burner, is an active stimulus to plant growth when used at night to supplement daylight.

2. Lettuce plants subjected to the influence of the incandescent gas light at night were taller and heavier than plants of the same variety and seed sowing grown in normal conditions.

3. That lettuce and spinach subjected to the stimulating influence of the light grew faster and completed their growth in less time than plants of the same sorts from the same seed sowing grown in normal conditions.

4. That no injurious effects resulted from the use of the incandescent gas light.

5. That the stimulating influence of the light as indicated by the growth of plants used in the various tests is shown by the order in which the sorts are named, the first being the most susceptible—spinach, cabbage, radish, lettuce, tomato.

6. That the range of the light is somewhat variable for the different crops. In general the maximum growth was attained at 12 to 16 feet from the light while a perceptible increase was noticed at 24 feet.

7. Bloom record of tomatoes shows markedly earlier bloom in the light house.—eight days the least and eighteen days the greatest difference.

8. That in the case of radishes top growth was stimulated but evidently not markedly at the expense of root. With

sugar beets top growth was greatly stimulated evidently at the expense of root growth.

9. That while the roots of beets grown in the normal house were larger than those in the light house, the sugar contents and the per cent. purity was markedly higher in the light house grown roots.

10. Spinach, lettuce and radishes all tend to make seed stalks earlier under the light than in normal conditions.

11. Lettuce and spinach under the influence of the incandescent gas light not only grew faster during the growing period but the period was actually longer than for plants in the normal house.—See auxanometer records, pages 88 and 89.

L. C. CORBETT.

NOTICE.

Application for bulletins of this Station should be addressed to the Director of the West Virginia Agricultural Experiment Station, Morgantown, W. Va.

(The bulletins named below are available for distribution.)

Special Bulletin No. 2. Proceedings connected with the cele-
bration upon the completion of the Station Building and
the organization of the Sheepbreeders and Wool-Growers'
Association and the State Horticultural Society.

Third Annual Report, 1890.

No. 51. Commercial Fertilizers, Jan., '98.

No. 52. Strawberries.

No. 53. Commercial Fertilizers, Dec. '98.

No. 54. Nursery Hints.

No. 55. Sugar Beets.

No. 56. Report on Investigations to Determine the Cause of
Unhealthy Conditions of the Spruce and Pine From 1880-
1893

No. 57. Commercial Fertilizers.

No. 58. The Effect of Pressure in the Preservation of Milk.

No. 59. Whole Corn Compared with Corn Meal for Fattening
Hogs.

No. 60. Poultry Experiments.

No. 61. Sheep Feeding Experiments.

No. 62. A Study of the Effect of Incandescent Gas-light on
Plant Growth.

www.ingramcontent.com/pod-product-compliance
Lightning Source LLC
Chambersburg PA
CBHW022014190326
41519CB00010B/1522